How to Dazzle at
ORAL AND MENTAL STARTERS

Beryl Webber and Jean Haigh

Brilliant Publications

We hope you and your class enjoy using this book. Other books in the series include:

Maths titles
How to Dazzle at Algebra   1 903853 12 5
How to Dazzle at Written Calculations   1 903853 11 7

English titles
How to Dazzle at Writing   1 897675 45 3
How to Dazzle at Reading   1 897675 44 5
How to Dazzle at Spelling   1 897675 47 X
How to Dazzle at Grammar   1 897675 46 1
How to Dazzle at Reading for Meaning   1 897675 51 8

Science title
How to Dazzle at Being a Scientist   1 897675 52 6

ICT title
How to Dazzle at Information Technology   1 897675 67 4

If you would like further information on these or other titles published by Brilliant Publications, please write to the address given below.

Published by Brilliant Publications, 1 Church View, Sparrow Hall Farm, Edlesborough, Dunstable, Bedfordshire LU6 2ES

Telephone: 01525 229720
Fax: 01525 229725

email: sales@brilliantpublications.co.uk
website: www.brilliantpublications.co.uk

Written by Beryl Webber and Jean Haigh
Illustrated by Sue Woollatt of Graham Cameron Illustrations
Cover illustrated by Darin Mount of Graham Cameron Illustrations

Printed in Malta by Interprint Limited

© Beryl Webber and Jean Haigh 2002
ISBN 1 903853 10 9

First published 2002
10 9 8 7 6 5 4 3 2 1

The right of Beryl Webber and Jean Haigh to be identified as the authors of this work has been asserted by them in accordance with the Copyright, Designs and Patents Act 1988.

Pages 6–48 may be photocopied by individual teachers for class use, without permission from the publisher. The material may not be reproduced in any other form or for any other purpose without the prior permission of the publisher.

Contents

	Page
Introduction	4
How to use the book	5
Halving	6
Money, money, money	7
Money bags	8
How much does it cost?	9
Shopping	10
Describing numbers	11
Vocabulary game	12
Number vocabulary	13
Digits	14
Counting tenths	15
Make 100	16
Make a number	17
Mid-points	18
Darts	19
How did I do it?	20
Finding pairs	21
Using factors	22
Multiplication grids	23
Finding percentages	24
Matching fractions	25
Matching fractions and decimals	26
Matching fractions and percentages	27
Ratios	28
Ratio and proportion	29
What's my average?	30
What's my weekly average?	31
Perimeters	32
Perimeters of shapes	33
Angle vocabulary	34
Transformation quick recall	35
Designs	36
Rotating shapes	37
Centimetres and millimetres	38
How long will my phone voucher last?	39
Telephone charges	40
Lunch for a week	41
Prices in a sale	42
Rapid response questions 1	43
Rapid response questions 2	45
Rapid response questions 3	47

Introduction

How to Dazzle at Oral and Mental Starters contains 43 photocopiable sheets for use with pupils aged 11 to 13 who are working at levels 2–3 of the National Curriculum. The activities are presented in an age-appropriate manner and provide a flexible but structured resource for teaching pupils to understand and use mental strategies for mathematical calculations. The tasks are varied but repeat these operations in a range of contexts and using different approaches. Each task is linked to the *National Numeracy Strategy Framework for Key Stage 3*. The tasks can be introduced to pupils in mixed-ability classes or are suitable to be undertaken by pupils working in the lower sets in Years 7 and 8.

These pupils will have had many years working on these mathematical concepts during their primary school education and it is important to ensure that fundamental principles of the number system are understood before embarking on the tasks. The tasks are intended to support direct teaching and give teachers some evidence to assess pupils' ability to use and adapt mental strategies. There is a wide range of strategies developed through the book in order that pupils will have opportunities to find a method that they understand and works for them.

These sheets are designed to give pupils opportunities to succeed in their mathematics. The expectation that the pupil will achieve successfully will help to build confidence and competence.

How to use the book

The activity pages are designed to supplement any numeracy programme you undertake in the classroom. They are intended to increase the pupils' oral and mental skills by giving opportunities to develop different mental approaches to arithmetic problems and to try different methods of working out mental mathematical problems.

The activities may be used by the teacher as oral and mental starter tasks or as support for individuals, pairs or groups of pupils who need more practice or require specific skills to enable them to develop their quick recall skills in a specific aspect of mathematics.

They can be used with individual pupils, pairs or very small groups, as the need arises. The mathematics is linked to the learning objectives for Year 7. Pupils with poor reading skills may need support with reading problems if the book is used to support the oral and mental starter activity. Often pupils experiencing learning difficulties have poor auditory memories. This needs to be taken into account when the sheets are used as oral activities to start a lesson. When the sheets are included as part of a follow-up to the oral and mental starter, teachers need to be aware of the poor visual memories of some pupils and ensure that the activity is understood and the context is familiar to the pupil. However, pupils should be able to extract the mathematics themselves and decide on the correct operation to apply.

There are three rapid response tests that should be used to identify strengths and weaknesses. They can be given to pupils as assessment exercises and the range of strategies used by individuals discussed during subsequent oral and mental starter sessions at the beginning of the lessons. Pupils could write their answers on a white board. There is no time limit for these tests. The prime purpose is to raise the self-esteem of the pupils and give them confidence that their accurate mental skills are improving.

It is not the intention of the authors that the teacher should expect all the pupils to complete all the sheets, rather that the sheets be used with a flexible approach, so that the book will provide a bank of resources that will meet the needs as they arise.

Many of the sheets can be modified and extended by creating further examples. The Add-ons provide a good vehicle for discussion of what has been learned and how it can be applied. The Add-ons should always be included in any class or group discussion at the end of the lesson, or in some cases may be suitable as homework tasks for discussion at a later date.

Halving

Find halfway. Write the number.

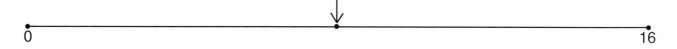

Now find halfway again.

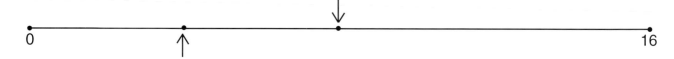

And again.

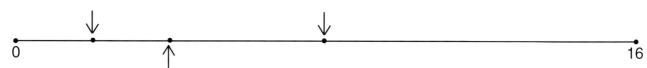

And again.

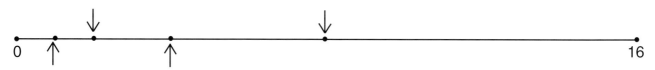

Try one more time.

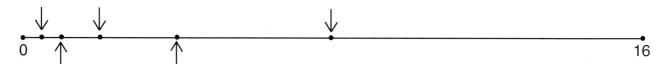

Add-on
Try finding halfway five more times.

Money, money, money

How many different amounts can you make from these coins?

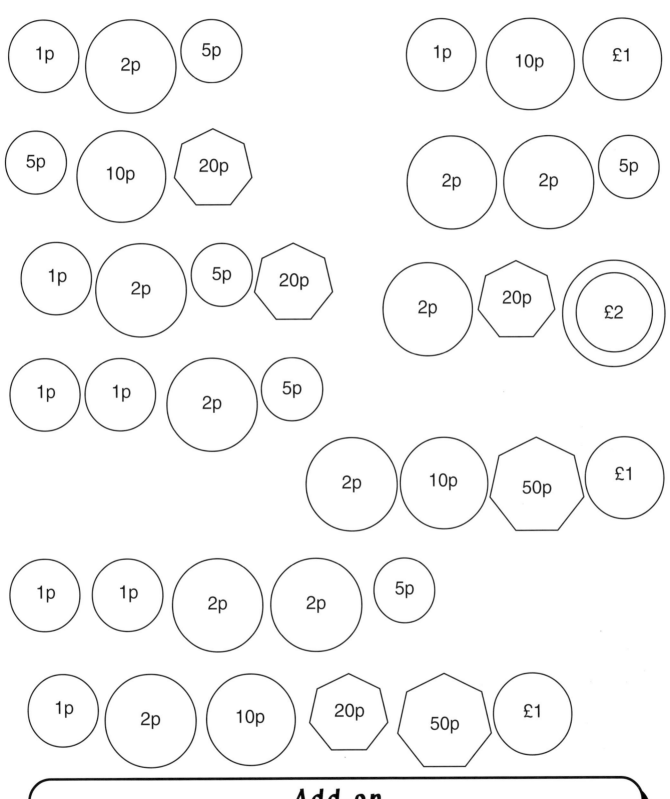

Add-on
What is the smallest number of coins you would need to be able to make all the amounts up to 50p? What about £1?

Money bags

In your bag you find some coins.
The total amount is £2.69.

What coins could you have found?

Make a grid to help you sort out as many combinations as possible that use fewer than twelve coins.

£2	£1	50p	20p	10p	5p	2p	1p

Add-on
How many ways can you make £7.75 using silver coins only?

How much does it cost?

A quick way of adding several numbers is to look for pairs that total a multiple of 10. For example: 15 and 25, 13 and 27, etc.

Try these:

36p + 23p + 37p + 14p =

18p + 94p + 12p + 56p =

44p + 19p + 26p + 21p =

82p + 18p + 34p + 76p + 55p + 25p =

42p + 29p + 28p + 53p + 71p + 47p =

62p + £1.06 + 27p + £3.18 + 94p + 33p =

8p + 76p + 12p + 34p + 25p + 15p =

18p + 24p + 13p + £1.02 + 96p + £1.07 =

11p + 17p + 50p + 50p + £1.03 + 99p =

£1.17 + £3.26 + £3.43 + 19p + 24p + 41p =

84p + 44p + £1.26 + 26p + 73p + 47p =

Add-on
Find a long supermarket bill. Cover the total amount at the bottom. See how quickly you can add it up.

Shopping

Many items in shops cost a bit less than the next pound or ten pounds:

To find out how much your shopping costs when you have lots of these amounts, round up to the next pound or ten pounds and then add. Finally subtract the right number of pence or pounds.

For example:

£9.99 + £7.99 = £18 − 2p = £17.98
↓   ↓
£10   £8

Try these for yourself:

49p + 69p =

79p + 99p =

69p + 59p + 99p =

19p + 39p + 79p + 99p =

£1.99 + £2.99 =

£3.99 + £9.99 =

£6.99 + £14.99 =

£7.99 + £6.99 + £8.99 =

£59 + £69 =

£99 + £19 + £49 + £59 =

Add-on
Discuss with a friend why shops price goods in this way.

Describing numbers

tenth	hundredth	thousandth
decimal number	less than	greater than
between	directed number	positive
negative	approximately	multiple
factor	divisible	prime

Use at least two of the words or phrases above to make up a sentence about each of these numbers.

81	
4.5	
−11	
99	
17	
6.02	
0.09	
4.792	
−15.7	
0.409	

Add-on
Read your sentences to a friend. See if they can guess which numbers you are describing.

Vocabulary game

Work as a team.

Share out the vocabulary cards.

Write a sentence and give an example for your card.

When you have finished all the cards compare your examples and sentences with another team.

share	remainder
multiple	product
factor	inverse

Number vocabulary

Reproduce the grid and list of words below on an overhead transparency. Divide the class into two teams. Ask them to match the words to the numbers in turn. Cross off the numbers as you go. The team to identify the most numbers wins.

$\frac{9}{8}$	45%	7	42	0.42
27	10	$1\frac{1}{3}$	3^2	1:2
−6	$\frac{4}{5}$	14	$0.\dot{3}$	81

pyramid	cubic	improper fraction
prime	fraction	directed number
percentage	mixed number	decimal
square	triangular	recurring decimal
index	ratio	rectangular

Digits

Place value means the position of a digit tells you its value.

For example:		251.648	
The	2	shows how many hundreds	200
The	5	shows how many tens	50
The	1	shows how many units	1
The	6	shows how many tenths	0.6
The	4	shows how many hundredths	0.04
The	8	shows how many thousandths	0.008

Make the largest number you can from these digits and a decimal point.

| 4 | 7 | 5 | 1 | • |

Now make the smallest number you can.

Make as many different numbers as you can.

Put them in order – smallest to largest.

Add-on
Use a calculator to total all your numbers.

Counting tenths

When adding numbers with tenths in your head choose the largest number and then use an imaginary number line to count on the next number.

0.9 + 2.7 =

Try these:

3.5 + 0.3	1.2 + 2.9	6.3 + 0.5
0.6 + 4.1	0.3 + 7.2	0.8 + 3.6
9.2 + 0.4	6.4 + 0.2	2.6 + 0.8

Now use the same method for the subtractions below, but remember to count backwards!

2.6 – 0.8	5.6 – 0.4	3.9 – 0.2
6.9 – 0.9	4.7 – 0.1	9.2 – 0.6
8.4 – 0.7	5.2 – 1.1	4.3 – 0.6

Try to do these as quickly as possible:

0.9 + 3.7	4.2 – 0.5	6.4 + 0.8
7.2 + 0.7	8.6 – 1.2	9.2 – 0.7
11.4 + 1.5	12.9 – 1.7	8.4 – 0.5
2.4 + 0.6	3.4 + 0.7	5.4 – 1.5

Add-on

Try these – but note the answers are negative!

0.9 – 3.4   1.6 – 2.1   3.4 – 5.7   2.5 – 3.1   0.5 – 0.9

Make 100

Choose 4 digits, say: 1   3   5   8

You can multiply any of the digits by 10.

You can add or subtract any numbers you make.

Try to get as close to 100 as possible.

For example:	
1. 8 × 10 = 80	4. 3 × 10 = 30
2. 1 × 10 = 10	5. 70 + 30 = 100
3. 80 − 10 = 70	

Only use each digit once.

Try this for yourself. Work with a friend.

Are there any sets of four digits that you cannot get close to 100?

Try with five digits.

Add-on
Try the same activity but you can multiply and divide your numbers now as well as adding and subtracting.

Make a number

Throw a die to choose five numbers between 1 and 6.

Use the numbers to make a number as close as possible to 500.

1. You can multiply any number by 5 or 10.

2. You can then add or subtract any numbers you have made.

Compare your answers with a friend.

Add-on
Try making 1000 with the same numbers.

Mid-points

Find the numbers that are halfway between these numbers.

For example:

```
    ↓
0   5   10
```

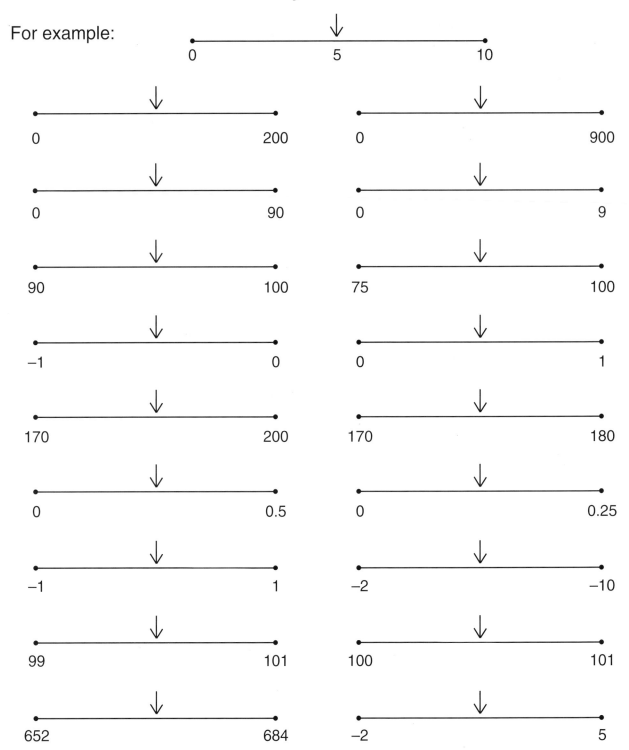

0 — 200	0 — 900
0 — 90	0 — 9
90 — 100	75 — 100
−1 — 0	0 — 1
170 — 200	170 — 180
0 — 0.5	0 — 0.25
−1 — 1	−2 — −10
99 — 101	100 — 101
652 — 684	−2 — 5

Add-on
Make up some more for a friend to try.

Darts

You throw three darts.

What is the smallest score you can get?

What is the largest?

What scores can you get that are:
- even?
- multiples of 3?
- multiples of 5?
- multiples of 10?

What scores can you get that are:
- prime?
- square?
- triangular?

How did I do it?

Here are some sums to work out in your head.

Explain how you worked out the answer.

How many different ways have you used?

| 59 + 74 | 193 + 29 | 3.46 + 4.59 |

| 101 − 84 | 1725 − 130 | 10.5 − 0.6 |

| 18 x 11 | 37 x 100 | 99 x 9 |

| 48 ÷ 4 | 1.25 ÷ 0.25 | 1800 ÷ 9 |

Add-on
Make up some of your own using 100 to
add, subtract, multiply and divide.
Are there any easy ways to work out the answers?

Finding pairs

Look at the grid below. Find pairs of numbers that add to make a whole number. For example: 3.6 + 1.4 = 5.0.

**Cross out each pair as you find them.
Which number is the odd one out?**

1.2	2.7	0.8	6.5	5.3
8.3	5.5	3.3	4.1	2.2
0.2	3.2	8.9	6.7	2.4
1.5	5.9	6.8	9.2	6.8
1.8	3.8	0.1	2.5	5.7

Add-on
Make a similar grid for a friend to try.
Make a similar grid for pairs of numbers that total 100.

Using factors

You can make multiplication more simple by breaking one of the numbers into its factors.

For example:

```
         12  x  18
                / \
               3   6
```

12 x 3 x 6 = 36 x 6 = 216

Try these:

12 x 15	12 x 20
12 x 16	12 x 21
12 x 24	12 x 30
13 x 24	13 x 30
17 x 36	17 x 24
15 x 19	20 x 27
16 x 32	18 x 17

Use a calculator to check your answers.

Add-on

Can you use a similar method to help you divide?

Multiplication grids

Fill in the missing numbers:

x	2	
	8	
10		60

x		5
7	21	
		20

x	5		8
			16
10		60	
		30	

x	4		9
	24		
10		75	
			13.5

x			5		9
0.1		0.2			
			4.5		
0.6	3			1.2	
			100		

Add-on
Make up some grids for a friend to try.

Finding percentages

> Example:
> Find 20% of £13
> First find 10% (£13 ÷ 10 = £1.30)
> Now multiply by 2 (£1.30 x 2 = £2.60)
> ➡ 20% of £13 = £2.60

Note: Remember 10% = $\frac{1}{10}$.

Try these:

10% of £17	20% of £17
30% of £15	40% of £13
20% of £22	50% of £25
80% of £16	40% of £18

> Example:
> Find 5% of £13
> First find 10% (£13 ÷ 10 = £1.30)
> Now divide by 2 (£1.30 ÷ 2 = £0.65)
> ➡ 5% of £13 = £0.65 (or 65p)

Try these:

5% of £17	5% of £22
5% of £18	5% of £36
15% of £17	15% of £22
35% of £24	45% of £19

Add-on

Think about how to find $2\frac{1}{2}$%. Use this information to calculate the VAT (at $17\frac{1}{2}$%) on different amounts.

Matching fractions

Find pairs of fractions that are equivalent.

| $\frac{1}{2}$ | $\frac{6}{8}$ | $\frac{2}{6}$ |

| $\frac{2}{3}$ | $\frac{1}{3}$ | $\frac{4}{16}$ |

| $\frac{1}{5}$ | $\frac{1}{4}$ | $\frac{8}{12}$ |

| $\frac{3}{5}$ | $\frac{4}{5}$ | $\frac{5}{10}$ |

| $\frac{3}{4}$ | $\frac{16}{20}$ | $\frac{2}{10}$ |

Which is the odd one out?

Add-on
Find another fraction that is equivalent to each pair.

Matching fractions and decimals

Match each fraction to a decimal number.

| $\frac{1}{4}$ | $\frac{3}{4}$ | $\frac{1}{3}$ |

| 0.2 | $\frac{6}{10}$ | 0.5 |

| 0.6 | $\frac{1}{2}$ | $\frac{2}{5}$ |

| $0.\dot{3}$ | 0.4 | 0.25 |

| $\frac{8}{10}$ | $\frac{1}{5}$ | 0.75 |

Which is the odd one out?

Add-on
Find another fraction that goes with each pair.

Matching fractions and percentages

Match each fraction to the equivalent percentage.

$\frac{90}{100}$	$\frac{3}{4}$	25%
$\frac{7}{10}$	50%	90%
30%	$\frac{1}{5}$	35%
$\frac{1}{2}$	$\frac{3}{12}$	20%
75%	$\frac{7}{20}$	$\frac{3}{10}$

Which is the odd one out?

Add-on
Find the decimal number that goes with each pair.

Ratios

Ratio compares part to part.

one part grey to two parts white
1:2

Write the ratio of the grey to white parts.

Make up two of your own.

Add-on

A bag has 36 marbles: 12 green, 18 red and 6 blue. What is the ratio? Write the ratio in its simplest form.

Ratio and proportion

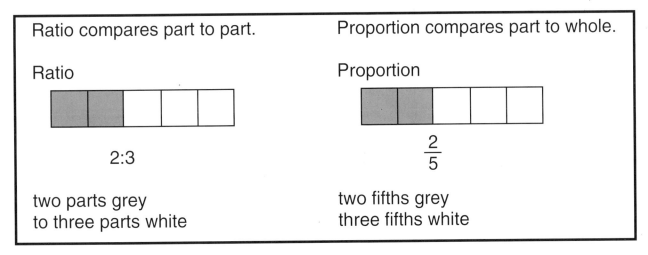

Ratio compares part to part.   Proportion compares part to whole.

Ratio   Proportion

2:3 $\frac{2}{5}$

two parts grey   two fifths grey
to three parts white   three fifths white

                              Ratio   Proportion

What is the connection between ratio and proportion?

Add-on
Proportion can be written as a fraction, percentage or decimal.
Write these ratios as proportions: 2:3, 5:8, 7:10.

What's my average?

The mean average is worked out by adding the information together and dividing by the total number of the different pieces of information.

How many sold?

	Monday	Tuesday	Mean average
Ice-creams	15	23	15 + 23 = 38 for 2 days The mean average for 2 days is 19 ice creams

Work out these mean averages:

	Wednesday	Thursday	Mean average
Ice-creams	32	28	
Lollies	57	43	
Choc ices	15	19	
Drinks	82	76	

Add-on
Calculate the mean average for ice-cream sales for all four days. How will you work it out?

What's my weekly average?

My lunches cost a different amount each day for a week.
What is the mean average?

Monday	£1.20
Tuesday	£2.45
Wednesday	£2.28
Thursday	£3.05
Friday	£1.92

The mean average is £ ☐

If I have £2.50 per day will I have enough money for my lunch on Thursday?

How much money will I have left on Friday afternoon?

If lunches cost 10% more, how much money would I need each day?

Add-on
If I have £3.00 per day and have spent £2.54 on Monday and £1.87 on Tuesday. How much can I spend on Wednesday?

Perimeters

Here are the perimeters of some rectangles.

Work out some possible lengths and widths for each of the rectangles.

Length	Width	Perimeter
		30 cm
		56 cm
		48 cm
		240 mm
		720 mm
		840 mm

Were your dimensions the same as your friends?

Add-on
Work out as many different sized rectangles that are possible with a perimeter of 360 mm (use whole numbers only).

Perimeters of shapes

What are the perimeters of these shapes in millimetres?

Write the answers inside the shapes.

Hint: 1 cm equals 10 mm.

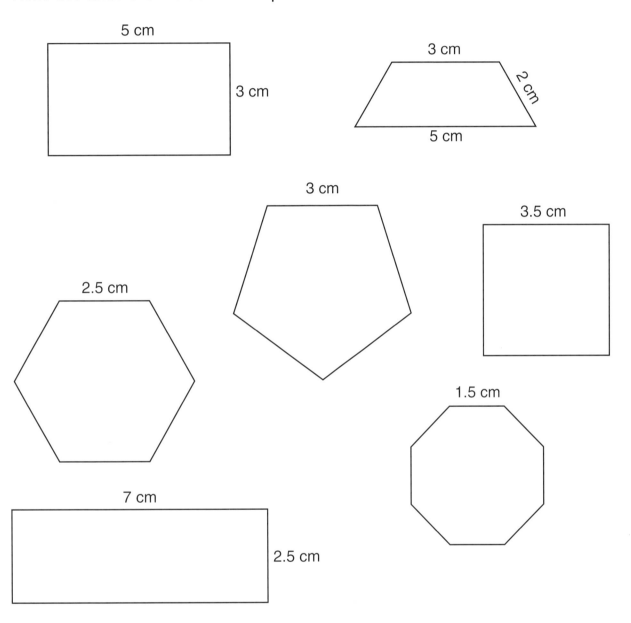

Hint: Regular polygons have sides that are all the same length.

Not all these shapes are regular polygons.

Add-on
Name all these shapes.

© Beryl Webber and Jean Haigh
This page may be photocopied for use by the purchasing institution only.

Angle vocabulary

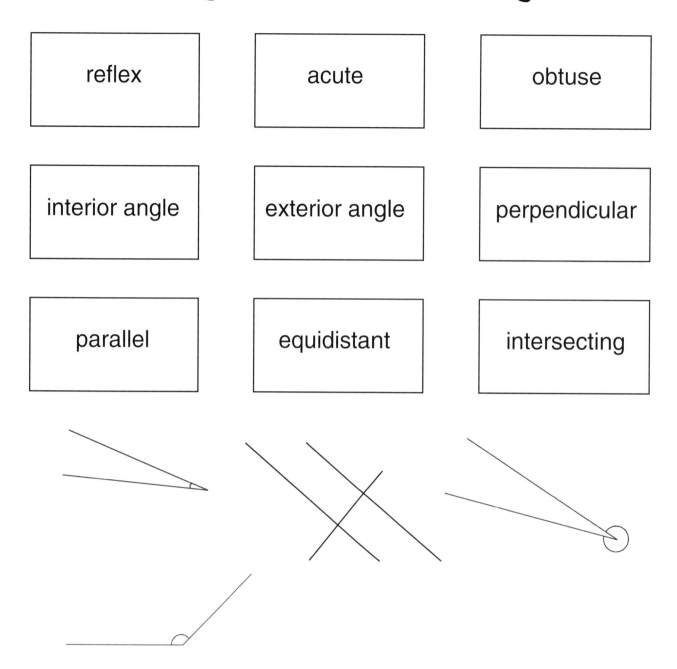

Which words or phrases can be used to describe these diagrams?

Make up a sentence using at least two of the words or phrases to describe each diagram.

Add-on
Make up some more sentences using at least two of the words or phrases to describe each diagram.

Transformation quick recall

You will need representations of the following shapes.

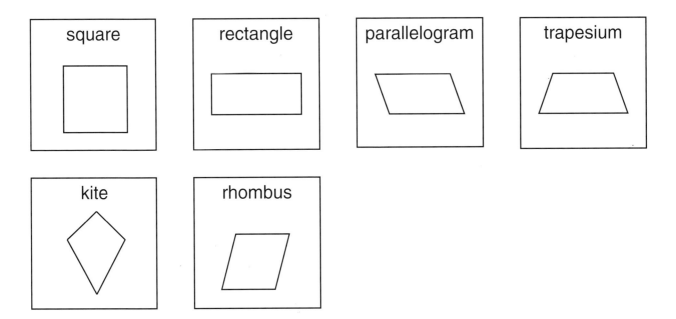

Choose a shape.

Rotate it 90°. Does it look the same?

Sort the shapes into those that look the same and those that don't.

Repeat, turning 180° and 270°.

Add-on
List the shapes with the order 2 of rotational symmetry.
What about turning triangles?

© Beryl Webber and Jean Haigh
This page may be photocopied for use by the purchasing institution only.

How to Dazzle at Oral and Mental Starters

Designs

Patterns can be made by changing the position of a shape.
The shape can be:

translated –
slid in any direction

reflected –
flipped over

rotated –
turned around

Look at these patterns and decide whether they are translated, reflected or rotated. Could they be all three?

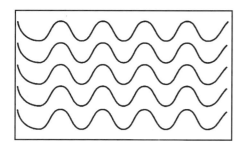  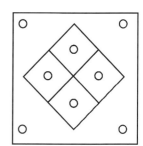

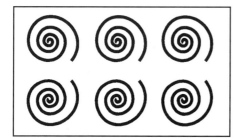  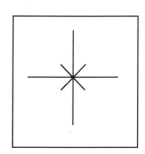

Add-on
Design your own pattern by translating, reflecting or rotating.

Rotating shapes

Here are some different shapes.

They can be turned from their centre points.

Reminder:
A shape has rotational symmetry if it looks the same two or more times in a full turn.

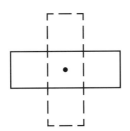

How many times can each of the shapes be turned so that it matches exactly the original position?

The rectangle can be moved to two positions.

This is called the order of rotational symmetry. A rectangle has a rotation of the order 2.

Try these shapes:

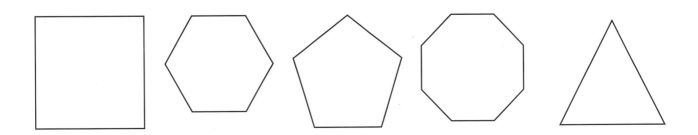

Write the order of rotational symmetry inside each shape.

Add-on
What is the order of rotational symmetry for these shapes?

isosceles triangle   parallelogram   trapesium

Centimetres and millimetres

10 millimetres (mm) equals 1 centimetre (cm).

Here are some 10 cm lines divided into two parts.

|———————•————————————————————|
       4 cm

What is the length of the 10 cm line in millimetres? ☐

What are the two parts of the line measured in millimetres?

☐ and ☐

Work out the two sections of these lines in millimetres.

|————•————————————————•————|
              7 cm

|——•————————————————————|
  2 cm

|————————————•——————————|
      6 cm

|————————————————•——————|
        8 cm

|——•————————————————————|
 1 cm

|———•———————————————————|
 1.5 cm

Reminder:
100 centimetres (cm)
equals 1 metre (m)

Add-on
How many millimetres are in 20 cm, 30 cm, 40 cm, 50 cm, 60 cm, 80 cm, 90 cm, 100 cm?

How long will my phone voucher last?

I have a £10.00 phone voucher to spend in one week.

These are my calls for one week.

	Sun	Mon	Tues	Wed	Thurs	Fri	Sat
No. of txt msgs	4	1	3	10	0	2	1
Peak calls			2 mins			5 mins	
Off-peak calls	5 mins			4 mins		7 mins	

Look at the table of charges below.

How much have I spent?

What other calls could I make?

Charges

Txt msgs	10p per msg
Peak calls	40p per minute
Off-peak	15p per minute

Add-on
Investigate ways of spending £5 in one week.

	Sun	Mon	Tues	Wed	Thurs	Fri	Sat
No. of txt msgs							
Peak calls							
Off-peak calls							

© Beryl Webber and Jean Haigh

Telephone charges

The telephone tariff has been changed.

Calls are now calculated by the second.

Peak	1p per second
Off-peak	0.5p per second

How much do these calls cost?

Peak rate

Time	57 secs	2 mins	5 mins	98 secs	24 secs	2 mins, 25 secs	4.5 mins	1 min, 22 secs
Cost								

Off peak-rate

Time	102 secs	80 secs	4 mins	2 mins, 25 secs	10 mins	5 mins, 18 secs	$\frac{1}{2}$ min	$\frac{1}{2}$ hour	15 mins
Cost									

Lunch for a week

You have £3.00 a day for lunch.

Plan your food bill for a week.

How much money can you save?

Choose some food for every day in one week.

Price list

Sandwiches	£1.50	Salad bar	£2.25
Hot meals	£2.00	Snacks	25p
		Drinks	50p

	Mon	Tues	Wed	Thurs	Fri	Weekly total
Sandwiches						
Hot meals						
Salad bar						
Snacks						
Drinks						
Daily total						

How much did you spend? ☐

How much did you save? ☐

Prices in a sale

Sale, Sale, Sale
Unbeatable prices
10% off CDs
Original price £12.00
Now only ☐

Sale, Sale, Sale
Unbeatable prices
12% off CDs
Original price £12.00
Now only ☐

Sale, Sale, Sale
Unbeatable prices
10% off computer games
Original price £35.00
Now only ☐

Sale, Sale, Sale
Unbeatable prices
12% off computer games
Original price £35.00
Now only ☐

Sale, Sale, Sale
Unbeatable prices
10% off T-shirts
Original price £15.00
Now only ☐

Sale, Sale, Sale
Unbeatable prices
12% off T-shirts
Original price £15.00
Now only ☐

Add-on
Prices are slashed again. How much would these items cost if there was a 15% discount?

Rapid response questions 1

(The numbers can be written on a white board.)

1. What is 15 more than 58?
2. What is the next prime number after 11?
3. 8 squared is ...
4. Double 217.
5. Halve 2510.
6. 5 x 8 = 40, what is 15 x 8?
7. Subtract 258 from 517.
8. Put these three fractions in ascending order:

    $\frac{1}{3}$ $\frac{5}{6}$ $\frac{3}{8}$

9. Write $\frac{3}{4}$ as a percentage.
10. Which of the following is a square number: 24, 32, 49, 54?

11. What would 2.15 pm be on a 24 hour clock?
12. Write $\frac{2}{5}$ as a decimal fraction.
13. Which of the following is divisible by 7: 17, 27, 34, 56, 72?
14. What is a quarter of 100?
15. Round 22,780 to the nearest thousand.
16. Divide 72 by 9.
17. Add 1.5, 1.25 and 3.00 together.
18. Take away a half from three-quarters.
19. Add £3.60, £4.10 and £1.25 together.
20. What is the inverse of 7 + 3 = 10?

Rapid response questions 2

(The numbers can be written on a white board)

1. My wages were increased from £23.75 to £25.15. What was the increase?
2. What is the next number in this sequence: 5, 2, −1, −4?
3. How many centimetres are there in 2.5 metres?
4. Halve 797.
5. What is the product of 7 and 9?
6. Round this number to the nearest ten thousand: 31,872.
7. Change these fractions to decimal fractions:
   $$\frac{1}{4} \quad \frac{3}{5} \quad \frac{1}{2}$$
8. What is the difference between £8.55 and £15.76?
9. Factorise 15.
10. Write the amount that is ten times more than £8.50.

11. Divide 54 by 7. What is the remainder?

12. What is halfway between 1.8 and 3.2?

13. Share equally 525 pencils between 5 people.

14. Simplify $\frac{3}{6}$.

15. A camera costing £150 is reduced by 50%. How much does it cost?

16. How many millimetres are there in 25 centimetres?

17. I bought a pair of shoes for £37.00. I paid with two £20 notes. How much change did I get?

18. I caught a train at 9.05 and arrived at 10.18. How long was the journey?

19. What is the average of these six temperatures: 15°C, 19°C, 18°C, 21°C, 22°C, 25°C?

20. I bought a game for £15. I sold it to my friend for £11.75. How much money did I lose?

Rapid response questions 3

(The number can be written on a white board.)

1. Write the number one hundred and eighty-three point 35 in figures.

2. One CD costs £12.99. What is the cost of five CDs?

3. I collected 20p pieces. When I counted them I had £15.60. How many 20p coins had I collected?

4. There are 30 students in a class. 20 are male. What fraction of the class is female?

5. What number is 57 more than 95?

6. There are 20 biscuits in a packet. 25% of them are wafers. How many wafer biscuits are in the packet?

7. What is a quarter of 96?

8. Multiply seven by eight and add six.

9. Subtract 3.85 from 9.05.

10. Add together 51, 102 and 1007.

11. Write 0.3 as a fraction.
12. Double 85 and add 35.
13. What is 57 divided by 10?
14. What is 57 multiplied by 10?
15. What is half of 1.62?
16. What is double 1.62?
17. What are the angles of an equilateral triangle? (All the angles of a triangle add up to 180°.)
18. Put these lengths in order, shortest to longest: 170 cm, 1.5 metres, 1600 millimetres.
19. Six people share £36.30 equally. How much do they each get?
20. My bus is due at 8.15 am. It takes 20 minutes to walk to the bus stop. What time do I need to leave home?